Table of Contents

PREFACE

On March 28, 2014 the Obama Administration released a key element called for in the President's Climate Action Plan: a Strategy to Reduce Methane Emissions. The strategy summarizes the sources of methane emissions, commits to new steps to cut emissions of this potent greenhouse gas, and outlines the Administration's efforts to improve the measurement of these emissions. The strategy builds on progress to date and takes steps to further cut methane emissions from several sectors, including the oil and natural gas sector.

This technical white paper is one of those steps. The paper, along with four others, focuses on potentially significant sources of methane and volatile organic compounds (VOCs) in the oil and gas sector, covering emissions and mitigation techniques for both pollutants. The Agency is seeking input from independent experts, along with data and technical information from the public. The EPA will use these technical documents to solidify our understanding of these potentially significant sources, which will allow us to fully evaluate the range of options for cost-effectively cutting VOC and methane waste and emissions.

1.0 INTRODUCTION

The oil and natural gas exploration and production industry in the U.S. is highly dynamic and growing rapidly. Consequently, the number of wells in service and the potential for greater air emissions from oil and natural gas sources is also growing. There were an estimated 504,000 producing gas wells in the U.S. in 2011 (U.S. EIA, 2012a), and an estimated 536,000 producing oil wells in the U.S. in 2011 (U.S. EIA, 2012b). It is anticipated that the number of gas and oil wells will continue to increase substantially in the future because of the continued and expanding use of horizontal drilling combined with hydraulic fracturing (referred to here as simply hydraulic fracturing) which allows for drilling in formerly inaccessible formations.

Due to the growth of this sector and the potential for increased air emissions, it is important that the U.S. Environmental Protection Agency (EPA) obtain a clear and accurate understanding of emerging data on air emissions and available mitigation options. This paper presents the Agency's understanding of air emissions and available control technologies from a potentially significant source of emissions in the oil and natural gas sector.

Oil and gas production from unconventional formations such as shale deposits or plays has grown rapidly over the last decade. Oil and natural gas production is projected to steadily increase over the next two decades. Specifically, natural gas development is expected to increase by 44% from 2011 through 2040 (U.S. EIA, 2013b) and crude oil and natural gas liquids (NGL) are projected to increase by approximately 25% through 2019 (U.S. EIA, 2013b). The projected growth of natural gas production is primarily led by the increased development of shale gas, tight gas, and coalbed methane resources utilizing new production technology and techniques such as horizontal drilling and hydraulic fracturing. According to the U.S. Energy Information Administration (EIA), over half of new oil wells drilled co-produce natural gas (U.S. EIA, 2013a). Based on this increased oil and gas development, and the fact that half of new oil wells co-produce natural gas, the potential exists for increased air emissions from these operations.

One of the activities identified as a potential source of emissions to the atmosphere during oil development is hydraulically fractured oil well completions. Completion operations are conducted to either bring a new oil well into the production phase, or to maintain or increase the well's production capability. Although the term "recompletion" is sometimes used to refer to completions associated with refracturing of existing wells, this paper will use the term "completion" for both newly fractured wells and refractured wells. In addition, hydraulically fractured coproducing oil wells can generate emissions of associated gas during the production phase. These processes and emissions are described in detail in Section 2.

The purpose of this paper is to summarize the EPA's understanding of VOC and methane emissions from hydraulically fractured oil well completions and associated gas during ongoing production. It also presents the EPA's understanding of mitigation techniques (practices and equipment) available to reduce these emissions, including the efficacy and cost of the technologies and the prevalence of use in the industry.

2.0 DEFINITION OF THE SOURCE

2.1 Oil Well Completions

For the purposes of this paper, a well completion is defined to mean:

The process that allows for the flowback of petroleum or natural gas from newly drilled wells to expel drilling and reservoir fluids and tests the reservoir flow characteristics, which may vent produced hydrocarbons to the atmosphere via an open pit or tank.

Completion operations with hydraulic fracturing are conducted to either bring a new oil well into the production phase or to maintain or increase the well's production capability (sometimes referred to as a recompletion). Well completions with hydraulic fracturing include multiple steps after the well bore hole has reached the target depth. These steps include inserting and cementing-in well casing, perforating the casing at one or more producing horizons, and often hydraulically fracturing one or more zones in the reservoir to stimulate production. Surface components, including wellheads, pumps, dehydrators, separators, tanks, and are installed as necessary for production to begin.

3

For the purposes of this paper, hydraulic fracturing is defined to mean:

The process of directing pressurized fluids containing any combination of water, proppant, and any added chemicals to penetrate tight formations, such as shale or coal formations, that subsequently require high rate, extended flowback to expel fracture fluids and solids during completions.

Hydraulic fracturing is one technique for improving oil and gas production where the reservoir rock is fractured with very high pressure fluid, typically a water emulsion with a proppant (generally sand) that "props open" the fractures after fluid pressure is reduced.

Oil well completions with hydraulic fracturing can result in VOC and methane emissions, which occur when gas is vented to the atmosphere during flowback. The emissions are a result of the backflow[1] of the fracture fluids and reservoir gas at high volume and velocity necessary to lift excess proppant and fluids to the surface. This comingled fluid stream (containing produced oil, natural gas and water) flows from each drilled well to a respective vertical separator and heater/treater processing unit. Fluid may be heated to aid in separation of the oil and natural gas and produced water. Phase separation is the process of removing impurities from the hydrocarbon liquids and gas to meet sales delivery specifications for the oil and natural gas. Oil may go directly to a pipeline or be stored onsite for future transfer to a refinery. If infrastructure is present, produced gas can be metered to a sales pipeline. If infrastructure is not available, the produced gas is frequently sent to combustion devices for destruction (e.g., flares) or is vented to the atmosphere.

Recompletions are conducted to minimize the decline in production, to maintain production, or in some cases to increase production. When oil well recompletions using hydraulic fracturing are performed, the practice and sources of emissions are essentially the same as for new well completions involving hydraulic fracturing, except that surface gas collection

---

[1] Backflow is the phenomena created by pressure differences between zones in the borehole. If the wellbore pressure rises above the average pressure in any zone, backflow will occur (i.e., fluids will move back towards the borehole). In contrast, "flowback" is the term used in the industry to refer to the process of allowing fluids to flow from the well following a treatment, either in preparation for a subsequent phase of treatment or in preparation for cleanup and returning the well to production.(http://www.glossary.oilfield.slb.com/)

equipment may already be present at the wellhead after the initial fracture. However, the backflow velocity during refracturing will typically be too high for the normal wellhead equipment (separator, dehydrator, lease meter), while the production separator is not typically designed for separating sand.

2.2   Associated Gas

Associated gas is the term typically used for natural gas produced as a by-product of the production of crude oil. Industry publications typically refer to associated gas as gas that is co-produced with crude oil while the well is in the production phase and is vented directly to the atmosphere or is flared. One published definition for associated gas is "gaseous hydrocarbons occurring as a free-gas phase under original oil-reservoir conditions of temperature and pressure (also known as gas-cap gas)."[2] Therefore, associated gas can include gas that is produced during flowback associated with completion activities and gas that is emitted from equipment as part of normal operations, such as natural gas driven pneumatic controllers and storage vessels. However, in this paper, the term "associated gas emissions" refers to:

> Associated gas emissions from the production phase (i.e., excluding completion events and emissions from normal equipment operations) that could be captured and sold rather than being flared or vented to the atmosphere if the necessary pipeline and other infrastructure were available to take the gas to market.

3.0   EMISSIONS DATA AND EMISSIONS ESTIMATES – HYDRAULICALLY FRACTURED OIL WELL COMPLETIONS

For consistency in the review of the various data sources and studies and to better understand the data discussions presented below, this section presents an overview of the types of the emissions estimation processes and the data that have been used in a number of studies to estimate VOC and methane emissions from hydraulically fractured oil well completions and recompletions.

---

[2] McGraw-Hill Dictionary of Scientific & Technical Terms, 6E, Copyright © 2003 by the McGraw-Hill Companies, Inc.

1) For estimating source emissions:
   - Gas produced during completions of oil wells. Estimated. This type of data would provide natural gas or methane production volumes for a completion. The data may be estimated using well characteristics (e.g., flow rate, casing diameter, and casing pressure) and established emission factors.
   - Gas produced by the oil well annually/daily/monthly. Direct measure or estimated. This type of data would be similar to the gas produced during completions but would be related to ongoing production of associated gas from the well.
   - Gas composition. This data is typically composition results from laboratory analysis of the raw gas stream to determine methane and other hydrocarbon volume or weight percent for use in converting natural gas or methane emissions estimates to VOC.
   - Duration of completion cycle. Length of the completion process in days.
   - Use of control technology. Flares, reduced emissions completions (RECs), other control technology or none. This information indicates whether a control device or practice is used and, if possible, the amount of produced gas captured and controlled.
2) For estimating nationwide emissions:
   - Number of oil well completions conducted annually. This information requires identification of the number of oil wells conducting completions/recompletions annually.
   - Number of oil wells co-producing natural gas. This involves identifying the population of oil wells using a definition of oil well based on some production criteria.
   - Number of oil wells completions with emissions controls such as RECs or flaring.

There are several available data sources for the data elements described above. Because most of the available data were not collected specifically for the purpose of estimating emissions, each source has to be qualified to ensure that the data are being used appropriately. In characterizing the nationwide emissions, we analyzed several sources of data and qualify each source with respect to the different aspects of the emission estimation process. Therefore, in addition to describing the data source and any relevant results of analysis, this paper discusses the implications of the data and/or results of analysis of the data with respect to the quantity of data, quantity of emissions, scope of emissions estimates, geographic dispersion, and variability in data.

Lastly, methodologies used in the emission estimation process are described, such as a discussion of the methodology for deriving emission factors or for identifying national populations.

There is variation in the industry as to how oil wells and gas wells are defined. Some publications do not differentiate at all between them, while others use the amount of oil produced or a gas-to-oil ratio (GOR) threshold as a dividing line between a gas well and an oil well. This paper does not attempt to choose a specific definition of "oil well," but instead describes the definitions used in each study or data source. The intent of this section of the paper is to present the EPA's understanding of the available data and its usefulness in estimating VOC and methane emissions from this source.

3.1   Summary of Major Studies and Sources of Emissions Data

Given the potential for emissions from hydraulically fractured oil well completions, there have been several information collection efforts and studies conducted to estimate emissions and available emission control options. Studies have focused on completion emission estimates. Some of these studies are listed in Table 3-1, along with an indication of the type of information contained in the study (i.e., activity level, emissions data, and control options).

Table 3-1. Summary of Major Sources of Information and Data on Oil Well Completions

Name	Affiliation	Year of Report	Activity Factor	Uncontrolled/Controlled Emissions Data	Control Options Identified
Fort Berthold Federal Implementation Plan (U.S. EPA, 2012a)	U.S. Environmental Protection Agency	2012	Regional	Uncontrolled	X
ERG/ECR Contractor Analysis of HPDI® Data	U.S. Environmental Protection Agency	2013	Nationwide	Uncontrolled	X
Environmental Defense Fund Analysis of HPDI® Data (EDF, 2014)	Environmental Defense Fund	2014	Nationwide	Uncontrolled	-

Name	Affiliation	Year of Report	Activity Factor	Uncontrolled/Controlled Emissions Data	Control Options Identified
Measurements of Methane Emissions at Natural Gas Production Sites in the United States (Allen et al., 2013)	Multiple Affiliations, Academic and Private	2013	26 Completion Events	Both	-
Methane Leaks from North American Natural Gas Systems (Brandt et. al, 2014a and 2014b)	Multiple Affiliations	2013	Regional	Uncontrolled	-

Data for Petroleum and Natural Gas Systems collected under the EPA's Greenhouse Gas Reporting Program (GHGRP) or the EPA's Inventory of U.S. Greenhouse Emissions and Sinks (GHG Inventory), are not discussed in detail in this section. The GHGRP does not require reporting of vented emissions from hydraulically fractured oil well completions. The GHG Inventory estimates emissions from oil well completions, but does not distinguish between completions/recompletions of conventional wells and completions/recompletions of hydraulically fractured wells.

A more-detailed description of the data sources listed in Table 3-1 is presented in the following sections, including how the data may be used to estimate national VOC and methane emissions from oil well completion events.

3.2   Fort Berthold Federal Implementation Plan (FIP) – Analysis by EC/R (U.S. EPA) 2012a)

On March 22, 2013, the EPA published (78 FR 17836) the FIP for existing, new and modified oil and natural gas production facilities on the Fort Berthold Indian Reservation (FBIR). In support of that effort, the EPA conducted an analysis of 154 applications for synthetic minor New Source Review (NSR) permits that indicated VOC emissions were the most prevalent of the pollutants emitted from the oil and natural gas production sources operating on the FBIR, which contain equipment that handles natural gas produced during well completions, phase separation during production, and temporary storage of crude oil (U.S. EPA, 2012).

The EPA FIP established federally enforceable requirements to control VOC emissions from oil and natural gas production activities that were previously unregulated or regulated less strictly. The FIP requires a 90%-98% reduction of VOC emissions from gas not sent to a sales line using pit flares, utility flares and enclosed combustors, all technologies which were found to be standard industry practice on the FBIR. The analysis included a large dataset of combustion control equipment cost information based on three well/control configuration scenarios.

The FBIR dataset includes:

- 533 production wells from five major operators
- Average controlled and uncontrolled VOC emissions from oil wells for wellhead gas, heater/treaters, and storage tanks
- Oil production data
- Number of sources; storage tanks, combustors, flares, and if a pipeline is present
- Current capital and annualized cost estimates for combustion and REC control options
- Gas composition data (for each permit application)
- Projected 2,000 new wells or 1,000 well pads per year between 2010 and 2029.

The data provided for the FBIR, although useful, has certain qualifying limitations. For instance, the FBIR data is primarily for wells producing from the Bakken and Three Forks formations, which limits it to a regional dataset. Also, the FBIR data showed high variability in oil well production rates and in product composition. This variability may not be representative of other formations. Also, according to the North Dakota Department of Health, the Bakken formation typically contains a high amount of lighter end VOC components which have the potential to produce increased volumes of flash emissions compared to typical oil production wells (U.S. EPA, 2012a). This may be somewhat unique to the Bakken formation and not be representative nationally.

Table 3-2 summarizes an analysis performed by EC/R of the FBIR data with respect to oil well completion emissions. The analysis estimated completion emissions by multiplying the average gas volume per day for each well by a 7 day flowback period. The analysis indicated that the average uncontrolled emissions from a well completion event are 37 tons of VOC per completion event.

Table 3-2. Summary of FBIR FIP Oil Well Completion Uncontrolled[3] Casing Gas and VOC Emissions

Data Element	Data from FBIR FIP											
	Enerplus	EOG	QEP[c]	WPX[b]	WPX-2[b]	WPX-3[b]	XTO[d]	Marathon	PetroHunt	Average	Min	Max
VOC Molecular weight	27.0	27.7	NA	28.1	29.6	31.7	24.5	28.5	25.8	27.8	24.5	31.7
Natural Gas Molecular weight	37.8	40.5	NA	43.7	45.9	51.0	32.9	41.4	34.3	41.0	32.9	51.0
Gas Constant (ft^3/lbmol)[a]	379	379	NA	379	379	379	379	379	379	379	379.0	379
Average Oil Production (bpd) - per well	1,181	255	NA	347	420	303	305	2,094	214	639.7	214	2,094
Average Gas Volume (Mcf/day) - per well	885	182	NA	250	292	210	305	491	197	351.5	182	885
Average Gas Volume (Mcf/completion)	6,197	1,272	NA	1,748	2,042	1,473	2,133	3,439	1,378	2,460	1,272	6,197
Average Uncontrolled VOC Emissions (ton/completion)	83	19	NA	28	37	31	23	53	16	37	16	83

NA = Not Reported, FBIR FIP = Fort Berthold Indian Reservation Federal Implementation Plan, EOG = EOG Resources, QEP = QEP Energy Co., WPX = WPX Energy, XTO = XTO Energy Inc.

a-Value used by North Dakota facilities represents 60°F and 1 atm. For subpart OOOO, this value is based on 68°F and 1 atm.

b-NOTE for WPX:

  i. They used three different molecular weights and percent. Therefore, each of these are represented in this table.

  ii. They only reported 10% of the VOC emissions because they flare 90% of their casinghead gas emissions. This table represents 100%.

c-The QEP molecular weight and VOC content data for casinghead gas were claimed as copyrighted and were not in the online docket.

d-XTO reported oil production and associated gas production as the same value. Therefore, did not include this gas to oil production ratio in the average.

[3] Uncontrolled emissions are the emissions that would occur if no emissions mitigation practices or technologies were used (e.g., completion combustion devices or RECs).

ERG Inc. and EC/R (ERG/ECR) conducted an analysis of Calendar Year (CY) 2011 HPDI[4] data to estimate uncontrolled emissions from hydraulically fractured oil well completions for the EPA. For this analysis the following methodology was used:

ERG extracted HDPI oil well data for hydraulically fractured, unconventional oil wells completed in CY 2011. Because the HPDI database does not differentiate between gas and oil wells, the following criteria were used to identify the population of hydraulically fractured oil well completions:

- Identified wells completed in 2011 using HPDI data covering U.S. oil and natural gas wells. Summary of the data and the logic for dates used is included in the memo "Hydraulically Fractured Oil Well Completions" (ERG, 2013)
- Identified wells completed in 2011 that were hydraulically fractured using the Department of Energy EIA formation type crosswalk supplemented with state data for horizontal wells (ERG, 2013)
- Determined which wells were oil wells based on their average gas-to-liquids ratio (less than 12,500 scf/barrel were considered to be oil wells)
- Estimated the average daily gas flow from the cumulative natural gas production for each well during its first 12 months of production
- The resulting dataset provided 192 data points representing county level average daily natural gas production at a total of 5,754 oil well completions for CY 2011.

Emissions in the ERG/ECR analysis were calculated using both a 3-day and a 7-day flowback period. The volume of natural gas emissions (in Mcf) per completion event was calculated using the average daily flow multiplied by both a 7-day flowback period and a 3-day flowback period. The gas volume was converted to mass of VOC using the same VOC

---

[4] HPDI, LLC is a private organization specializing in oil and gas data and statistical analysis. The HPDI database is focused on historical oil and gas production data and drilling permit data. For certain states and regions, this data was supplemented by state drilling information. The 2011 data was the most current data available when the analysis was performed.

composition and conversion methodology used for gas wells in the subpart OOOO well completion evaluation. The composition values used were 46.732% by volume of methane in natural gas and 0.8374 pound VOC per pound of methane for oil wells (EC/R, 2011a).

The analysis of the 2011 HPDI data for oil well completions provided an average gas production of 262 Mcf per well per day. Based on this gas production, the average uncontrolled VOC emissions were 20 tons per completion event based on a 7-day flowback period and 6.4 tons of VOC per completion event based on a 3-day flowback period. The average uncontrolled methane emissions were 24 tons per completion event based on a 7-day flowback period and 7.7 tons of methane per completion event based on a 3-day flowback period. It was assumed that the emissions for an oil well recompletion event are the same as an oil well completion event.

To estimate nationwide uncontrolled emissions for hydraulically fractured oil well completions, the average methane and VOC emissions per event were multiplied by the total number of estimated oil well completions. For 2011, which was the most recent data available in HPDI, the estimated nationwide uncontrolled hydraulically fractured oil well completion VOC emissions are 116,230 tons per year (i.e., VOC emissions/completion of 20.2 tons/event times the total oil well completion events per year of 5,274) based on a 7-day flowback period and 36,825 tons per year (i.e., VOC emissions/completion of 6.4 tons/event times the total oil well completion events per year of 5,274) based on a 3-day flowback period. The estimated nationwide uncontrolled hydraulically fractured oil well completion methane emissions are 138,096 tons per year (i.e., methane emissions/completion of 24 tons/event times the total oil well completion events per year of 5,274) based on a 7-day flowback period and 44,306 tons per year (i.e., VOC emissions/completion of 7.7 tons/event times the total oil well completion events per year of 5,274) based on a 3-day flowback period. Table 3-3 presents the results of the emission estimate analysis for both the 7-day and 3-day completion duration periods.

Table 3-3. Summary of Oil Well Completion Uncontrolled Emissions from 2011 HPDI Data

	7-day event	3-day event
Total number of hydraulically fractured oil well completions in 2011	5,754	5,754
Number of county well production averages (data points)	195	195
Natural Gas production per well, per day, weighted average (Mcf)	262	262
Methane emissions per completion/recompletion event, weighted average (tons)	24	7.7
VOC emissions per completion/recompletion event, weighted average (tons)	20.2	6.4
Uncontrolled Nationwide methane emissions, oil well completions (tpy)	138,096	44,306
Uncontrolled Nationwide VOC emissions, oil well completions (tpy)	116,230	36,825

Note: This estimate does not include recompletion emissions.

As stated earlier, these estimates are for uncontrolled emissions, thus estimates assume no control technology applied. National-level data on the prevalence of the use of RECs or combustors for reduction of emissions from oil well completion or recompletion operations were unavailable for this analysis.

State level information for Colorado, Texas and Wyoming on oil well recompletion counts was used to determine a percentage of producing wells for which recompletions were reported. The state level data were obtained for Colorado, Texas and Wyoming for recent years (COGCC, 2012, Booz, 2008 and RRCTX, 2013). Based on the state level data, it was determined that the average percentage of producing well undergoing recompletion was 0.5%. This includes both conventional and hydraulically fractured oil wells (the data did not allow the different types of wells to be distinguished from each other). Table 3.4 presents a summary of this analysis.

Table 3-4. Analysis of Texas, Wyoming and Colorado Recompletions Counts

State Data Source	Year	Total Number of Producing Wells	Total Number of Recompletions	Percent Recompletions to Total Producing Wells
Railroad Commission of Texas	2012	168,864	685	0.4
Wyoming Heritage Foundation	2007	37,350	304	0.8
State of Colorado Oil & Gas Conservation Commission	2012	50,500	152	0.3
Average Percent				**0.5**

While the state level recompletion data are recent, the percentage of producing oil wells that undergo recompletion in future years may increase due to more prevalent use of hydraulic fracturing on oil wells. However, no data have been obtained to quantify any potential increase in the oil well recompletion rate. This percentage was not used to estimate the number of recompletions of hydraulically fractured oil wells, because the data did not distinguish between conventional wells and hydraulically fractured wells.

3.4 Environmental Defense Fund and Stratus Consulting Analysis of Oil Well Completions[5] (EDF, 2014)

The Environmental Defense Fund (EDF) and Stratus Consulting (EDF/Stratus) conducted an analysis of HPDI data for oil wells to determine the cost effectiveness of the use of RECs and flares for control of oil well completion emissions within three major unconventional oil play formations, Bakken, Eagle Ford and Wattenberg. The oil well completion population was extracted using the DI Desktop for all oil wells with initial production in 2011 and 2012. Different filters were applied in each formation in order to identify the hydraulically fractured oil wells:

---

[5] This analysis is described in the EDF white paper "Co-Producing Wells as a Major Source of Methane Emissions: A Review of Recent Analyses" (http://blogs.edf.org/energyexchange/files/2014/03/EDF-Co-producing-Wells-Whitepaper.pdf). It is referred to in that paper as the "EDF/Stratus Analysis." The supplemental materials, including the data that was used in the analysis are available at https://www.dropbox.com/s/osrom4w6ewow4ua/EDF-Initial-Production-Cost-Effectiveness-Analysis.xlsx.

- Eagle Ford
  - Well Production Type: Oil
- Bakken
  - Well Production Type: Oil and Oil & Gas
- Wattenberg
  - Well Production Type: Oil

The resulting dataset included 3,694 oil wells for the Bakken formation, 1,797 oil wells for the Eagle Ford formation, and 3,967 oil wells for the Wattenberg formation. The assumptions EDF/Stratus made while conducting this analysis were:

- Well completions lasted an average of 7 to 10 days and the total gas production over that period was equal to 3 days of "Initial Gas Production" as reported in DI Desktop (i.e., 3 days of "Initial Gas Production" was equal to the uncontrolled natural gas emissions from the oil well completion).
- The natural gas content was 78.8% methane.

Table 3-5 summarizes the results of this analysis.

**Table 3-5. EDF Estimated Uncontrolled Methane Emissions
from Oil Well Completions Based on Analysis of HPDI® Oil Well Production Data**

Formation	Wells (#)	Uncontrolled Completion Emissions (gas Mcf/event)	Uncontrolled Completion Emissions (MT CH$_4$/event)	Uncontrolled Completion Emissions (tons CH$_4$/event)
Wattenberg[a]	3,967	624	9.5	10.5
Bakken[b]	3,694	1,183	18.0	19.8
Eagle Ford[c]	1,797	1,628	24.7	27.2

All results represent mean values.
a - Production data was downloaded for all oil wells in the Colorado Wattenberg formation with a first production date between 1/1/2010 and 3/1/2013.
b - Production data was downloaded for wells in the North Dakota Bakken formation with a completion date from 1/1/2010-12/31/2012. North Dakota does not distinguish between oil and gas wells. All wells with the type O&G were assumed to be oil wells.
c - Production data was downloaded for all oil wells in the Texas Eagle Ford formation with a completion date between 1/1/2010 and 2/23/2013.

The EDF/Stratus Analysis also provided an estimate of uncontrolled methane emissions from oil well completions of 247,000 MT (272,000 tons), however, the materials describing the analysis do not explain how this estimate was calculated.

3.5 Measurements of Methane Emissions at Natural Gas Production Sites in the United States (UT Study) (Allen et al., 2013)

The UT Study was primarily authored by University of Texas at Austin and was sponsored by the EDF and several companies in the oil and gas production industry. The study was conducted to gather methane emissions data at onshore natural gas well sites in the U.S. and compare the data to the EPA's Inventory of U.S. Greenhouse Gas Emissions and Sinks (GHG Inventory). The sources and operations that were tested included well completion flowbacks, well liquids unloading, pneumatic pumps and controllers and equipment leaks. The full study analysis included 190 onshore natural gas sites, which included 150 production sites, 26 well completion events, 9 well unloading and 4 well recompletions or workovers.

Six of the completion events in the UT Study were at co-producing wells (at least some oil was produced). The study reported the total oil produced, the total associated gas produced, the potential and actual methane emission, the completion duration, the type of emission control used, and the percent reduction from the control that was observed (Note: for two of the completion events, data was not gathered for the initial flow to the open tank). The data for these wells are summarized in Table 3-6.

Table 3-6. Summary of Completion Emissions from Co-Producing Wells

Site ID	Oil Produced (bbl)	Gas Produced (Mcf)	GOR (scf/bbl)	Potential Methane Emissions[a] (Mcf)	Actual Methane Emissions[b] (Mcf)	% Reduction	Data Analyzed	Duration (hrs)	REC or Flare
GC-1	1,594	6,449.9	4,046.36	5,005	106	97.9	Yes	75	Flare
GC-2	1,323	5,645	4,266.82	4,205	91	97.8	Yes	76	Flare
GC-3	2,395	26,363	11,007.52	21,500	264	98.8	Yes	28	REC
GC-4	1,682	24,353	14,478.60	13,000	180	98.6	Yes	28	REC
GC-6	448	13,755	30,703.13	12,150	247	98	No[d]	164	Flare
GC-7	1,543	5,413	3,508.10	4,320	90	97.9	No[d]	108	Flare

a – Measured emissions before flare or REC.
b - Measured emissions after flare or REC.
c - Calculated from measured before and after control.
d -Data not used in developing average emissions factor in the UT Study because, in these flowbacks, the study team was unable to collect completion emissions data for the initial flow to the open tank.

Using the threshold of a GOR of 12,500 scf/barrel to distinguish oil wells from gas wells, wells GC-1, GC-2, GC-3, and GC-7 would be considered oil wells. The average uncontrolled methane emissions from those wells were 213 tons (10,237 Mcf) and the average controlled (actual) emissions were 3.2 tons (154 Mcf).[6] The average duration of the completion for these wells was 72 hours (3 days). It is also worth noting that well GC-3 was controlled using a REC and 98.8% of the potential methane emissions were mitigated, demonstrating that RECs can be used effectively to control emissions from hydraulically fractured oil wells.

3.6   Methane Leaks from North American Natural Gas Systems (Brandt et. al, 2014a and 2014b)

Novim, a non-profit group at the University of California, sponsored a meta-analysis of the existing studies on emissions from the production and distribution of natural gas. As part of this analysis, Novim estimated emissions from hydraulically fractured oil well completions based on data from HPDI®. Novim included wells that were drilled in 2010 or 2011 in the Eagle Ford,

---

[6] These averages do not include well GC-7, because, as noted above, data from this well was not used in the UT Study due to the inability to collect all the emissions data.

Bakken, and Permian formations (Brandt et. al., 2014a). Different filters were applied in each formation in order to identify the hydraulically fractured oil wells:

- Eagle Ford
  - Well Production Type: Oil
  - Drill Type: Horizontal
- Bakken
  - Well Production Type: Oil and Oil & Gas
  - Drill Type: Horizontal
- Permian
  - Well Production Type: Oil
  - Drill Type: All

Using this method of qualifying the well population, Novim concluded 2,969 hydraulically fractured oil wells were completed in 2011 in the three formations (Brandt et. al., 2014a). In order to estimate completion emissions, Novim used the O'Sullivan method[7] in which peak gas production (normally the production during the first month) is converted to a daily rate of production. The O'Sullivan method assumes that during flowback emissions increase linearly over the first nine days until the peak rate is reached. Table 3-7 summarizes the estimated uncontrolled methane emissions per completion calculated by the Novim study.

Table 3-7. Summary of Uncontrolled Completion Emissions from Co-Producing Wells

Formation	Uncontrolled Methane Emissions (tonnes/event)[a]	Uncontrolled Methane Emissions (ton/event)[b]
Eagle Ford	90.9	93
Bakken	31.1	31.9
Permian	31.2	31.9

a – 1 Mg = 1 metric tonne of methane
b – Converted to U.S. short tons. 1 tonne = 1.02311 tons (short/U.S.) of methane

---

[7] O'Sullivan, Francis and Sergey Paltsev, "Shale gas production: potential versus actual greenhouse gas emissions", Environmental Research Letters, United Kingdom. November 26, 2012.

The Novim Study assumes methane emissions from these formations are representative of total national methane emissions from hydraulically fractured oil well completions and estimates those emissions to be 0.12 Tg (120,000 tonnes or 122,773 tons) per year for 2011.

It should be noted that the methodology in this study, like the ERG/ECR Analysis and the EDF/Stratus Analysis, uses gas production from HPDI® to estimate completion emissions. However, Novim uses the O'Sullivan method in which the emissions increase linearly through the flowback period until a peak is reached, while the ERG/ECR Analysis and the EDF/Stratus Analysis assume emissions are constant through the flowback period.

4.0 EMISSIONS DATA AND EMISSIONS ESTIMATES – ASSOCIATED GAS FROM HYDRAULICALLY FRACTURED OIL WELLS

Given the potential for emissions of associated gas from oil production, available information sources have been reviewed as to their potential use for characterizing the VOC and methane emission from associated gas production at oil well sites. As was stated previously, the term "associated gas emissions" in this paper refers to emissions from gas that is vented during the production phase that could otherwise be captured and sold if the necessary pipeline infrastructure was available to take the gas to market.

One methodology for estimating emissions would be to use the GOR of the well, which is a common piece of well data in the industry. An emission factor based on average GOR could be developed, and then the emission factor could be used to estimate uncontrolled associated gas emissions by applying it to known oil production (assuming all gas produced at an oil well is included in uncontrolled associated gas emissions). However, research indicates that associated gas production from oil wells declines over the life of the well, similar to oil production, but the decline is typically at a different rate than the oil production (EERC, 2013). This phenomenon introduces another variable into the analysis.

A second approach would be to use gas production reported for the well for economic and regulatory reasons. Conceivably, gas production could be used to estimate uncontrolled associated gas emissions. However, the EPA is not aware of a methodology that would allow the Agency to calculate the percentage of produced gas that could be captured if pipeline infrastructure were available. Some gas is emitted from equipment as part of normal operations, such as bleeding from pneumatic controllers. These emissions would not qualify as associated gas emissions as they have been defined in this paper.

The GHGRP does require reporting of "associated gas venting and flaring emissions." Additionally, the Ceres report contains data potentially useful for basic evaluation of VOC and methane associated gas emissions, but does not provide national estimates or per well estimates of emissions (Ceres, 2013). Both these sources are discussed in detail in the sections below.

The GHG Inventory does not include a category that specifically covers all associated gas emissions. Instead, these emissions are estimated in several categories in Petroleum Systems, and in Natural Gas Systems (emissions downstream of the gas-oil separator, and flaring).

4.1   Greenhouse Gas Reporting Program (U.S. EPA, 2013)

In October 2013, the EPA released 2012 greenhouse gas (GHG) data for Petroleum and Natural Gas Systems[8] collected under the GHGRP. The GHGRP, which was required by Congress in the FY2008 Consolidated Appropriations Act, requires facilities to report data from large emission sources across a range of industry sectors, as well as suppliers of certain GHGs and products that would emit GHGs if released or combusted.

When reviewing this data and comparing it to other datasets or published literature, it is important to understand the GHGRP reporting requirements and the impacts of these requirements on the reported data. The GHGRP covers a subset of national emissions from Petroleum and Natural Gas Systems; a facility[9] in the Petroleum and Natural Gas Systems source

---

[8] The implementing regulations of the Petroleum and Natural Gas Systems source category of the GHGRP are located at 40 CFR Part 98 Subpart W.

[9] In general, a "facility" for purposes of the GHGRP means all co-located emission sources that are commonly owned or operated. However, the GHGRP has developed a specialized facility definition for onshore production. For onshore production, the "facility" includes all emissions associated with wells owned or operated by a single company in a specific hydrocarbon producing basin (as defined by the geologic provinces published by the American Association of Petroleum Geologists).

category is required to submit annual reports if total emissions are 25,000 metric tons carbon dioxide equivalent (CO_2e) or more. Facilities use uniform methods prescribed by the EPA to calculate GHG emissions, such as direct measurement, engineering calculations, or emission factors derived from direct measurement. In some cases, facilities have a choice of calculation methods for an emission source.

Under the GHGRP, facilities report associated gas vented and flared emissions. Vented emissions are calculated based on GOR and the volume of oil produced and flared emissions using a continuous flow measurement device or engineering calculation. For 2012, 171 facilities reported associated gas vented and flared emissions to the GHGRP. Total reported methane emissions were 89,535 MT.

4.2   FLARING UP: North Dakota Natural Gas Flaring More Than Doubles in Two Years (Flaring Up) (CERES, 2013)

The Flaring Up report discusses the increase in North Dakota's oil and gas production from the Bakken formation between 2007 and mid-2013, the increased flaring of associated gas, and the potential value of NGL lost as a result of flaring. The report presents some associated gas production and flaring data that the authors derive from the gas production and flaring data reported by the North Dakota Industrial Commission (NDIC), Department of Mineral Resources. The Commission defines associated gas to be all natural gas and all other fluid hydrocarbons not defined as oil. Oil is defined by the Commission to be all crude petroleum oil and other hydrocarbons, regardless of gravity which are produced at the wellhead in liquid form and the liquid hydrocarbons known as distillate or condensate recovered or extracted from gas, other than gas produced in association with oil and commonly known as casinghead gas[10].

This Flaring Up report indicates that of the wells that are flaring the associated gas, approximately 55% are wells are not connected to a gas gathering system, while 45% are wells that are already connected. In addition, the report states that in May of 2013, 266,000 Mcf per day was flared, which represents nearly 30% of the gas produced (CERES, 2013). Percent flaring is currently reported by the NDIC while the connection data is tracked by the North Dakota

---

[10] North Dakota Century Code, Section I, Chapter 38-08 Control of Gas & Oil Resources, Section 38-08-02.

Pipeline Authority. The report concludes that the reason for the flaring of the associated gas is lack of pipeline infrastructure, lack of capacity and lack of compression infrastructure.

The data and information in this report is useful for discussion on the relative percentages of gas emissions being flared. The data, however, are specific to the Bakken, a formation that possesses unique characteristics both with regard to reservoir and formation characteristics, gas composition and the lack of infrastructure due to rapid development of the industry in the area.

5.0 AVAILABLE EMISSION MITIGATION TECHNIQUES

Two mitigation techniques were considered that have been proven in practice and in studies to reduce emissions from well completions and recompletions: REC and completion combustion. One of these techniques, REC, is an approach that not only reduces emissions but delivers natural gas product to the sales meter that would otherwise be vented. The second technique, completion combustion, destroys the organic compounds. Both of these techniques are discussed in the following sections, along with estimates of the efficacy at reducing emissions and costs for their application for a representative well. Combustion control for control of associated gas emissions (e.g., flaring) has been demonstrated as effective in the industry. However, flaring results in the destruction of a valuable resource and, as such, alternate uses for uncaptured/sold associated gas have been the subject of several studies with respect to new emerging technologies.

5.1 Reduced Emission Completions (REC)

5.1.1 Description

Reduced emissions completions are defined for the purposes of this paper as:

A well completion following fracturing or refracturing where gas flowback that is otherwise vented is captured, cleaned, and routed to the flow line or collection system, re-injected into the well or another well, used as an onsite fuel source, or used for other useful purpose that a purchased fuel or raw material would serve, with no direct release to the atmosphere.

Reduced emission completions, also referred to as "green" completions, use specially designed equipment at the well site to capture and treat gas so it can be directed to the sales line. This process prevents some natural gas from venting and results in additional economic benefit from the sale of captured gas and, if present, gas condensate. It is the EPA's understanding that the additional equipment required to conduct a REC may include additional tankage, special gas-liquid-sand separator traps and a gas dehydrator. In many cases, portable equipment used for RECs operates in tandem with the permanent equipment that will remain after well drilling is completed (EC/R, 2010b). In other instances, permanent equipment is designed (e.g., oversized) to specifically accommodate initial flowback. Some limitations exist for performing RECs because technical barriers vary from well to well. Three main limitations include the following:

- Proximity of pipelines. For certain wells, no nearby sales line may exist. The lack of a nearby sales line incurs higher capital outlay risk for exploration and production companies and/or pipeline companies constructing lines in exploratory fields.

- Pressure of produced gas. Based on experience using RECs at gas wells, the EPA understands that during each stage of the completion process, the pressure of flowback fluids may not be sufficient to overcome the sales line backpressure. In this case, combustion of flowback gas is one option, either for the duration of the flowback or until a point during flowback when the pressure increases to flow to the sales line.

- Inert gas concentration. Based on experience using RECs at gas wells, if the concentration of inert gas, such as nitrogen or carbon dioxide, in the flowback gas exceeds sales line concentration limits, venting or combustion of the flowback may be necessary for the duration of flowback or until the gas energy content increases to allow flow to the sales line. Further, since the energy content of the flowback gas may not be high enough to sustain a flame due to the presence of the inert gases, combustion of the flowback stream would require a continuous ignition source with its own separate fuel supply.

<u>5.1.2 Effectiveness</u>

Based on data available on RECs use at gas wells, the emission reductions from RECs can vary according to reservoir characteristics and other parameters including length of completion, number of fractured zones, pressure, gas composition, and fracturing technology/technique. Based on the results reported by four different Natural Gas STAR Partners who performed RECs primarily at natural gas wells, a representative control efficiency of 90% for RECs was estimated. The companies provided both recovered and total produced gas, allowing for the calculation of the percentage of the total gas which was recovered. This estimate was based on data for more than 12,000 well completions (ICF, 2011). Any amount of gas that cannot be recovered can be directed to a completion combustion device in order to achieve a minimum 95% reduction in emissions. Additionally, both wells that co-produced oil and gas and were controlled with a REC in the UT Austin study achieved greater than 98% reduction in methane emissions.

<u>5.1.3 Cost</u>

The discussion of cost in this section is based on the EPA's experience with RECs at gas wells. It is the EPA's understanding that the same equipment is used for RECs at gas wells and co-producing oil wells. All completions incur some costs to a company. Performing a REC will add to these costs. Equipment costs associated with RECs vary from well to well. High production rates may require larger equipment to perform the REC and will increase costs. If permanent equipment, such as a glycol dehydrator, is already installed or is planned to be in place at the well site as normal operations, costs may be reduced as this equipment can be used or resized rather than installing a portable dehydrator for temporary use during the completion. Some operators normally install equipment used in RECs, such as sand traps and three-phase separators, further reducing incremental REC costs.

The average cost of RECs was obtained from data shown in the Natural Gas STAR Lessons Learned document titled "Reduced Emissions Completions for Hydraulically Fractured Natural Gas Wells" (U.S. EPA, 2011a). The impacts calculations use the cost per day for gas capture and the duration of gas capture along with a setup/takedown/transport cost and a flare cost to represent the total cost. The cost is then annualized across the time horizon under study.

Costs of performing a REC are projected to be between $700 and $6,500 per day (U.S. EPA, 2011a). This cost range is the incremental cost of performing a REC over a completion without a REC, where typically the gas is vented or combusted because there is an absence of REC equipment. These cost estimates are based on the state of the industry in 2006 (adjusted to 2008 U.S. dollars). [11] Cost data used in this analysis are qualified below:

- $700 per day (equivalent to $806 per day in 2008 dollars) represents completion and recompletion costs where key pieces of equipment, such as a dehydrator or three-phase separator, are already found onsite and are of suitable design and capacity for use during flowback.
- $6,500 per day (equivalent to $7,486 in 2008 dollars) represents situations where key pieces of equipment, such as a dehydrator or three-phase separator, are temporarily brought onsite and then relocated after the completion.

The average of the above data results in an average incremental cost for a REC of $4,146 per day (2008 dollars).[12] The total cost of the REC depends on the length of the flowback period, and thus the length of the completion process. For example, if the completion takes 7 days then the total cost would be $29,022, and if the completion takes 3 days then the total cost would be $12,438 versus an uncontrolled completion. These costs would be mitigated by the value of the captured gas. The extent of this cost mitigation would depend on the price of the gas and the quantity that was captured during the REC.

---

[11] The Chemical Engineering Cost Index was used to convert dollar years. For REC, the 2008 value equals 575.4 and the 2006 value equals 499.6.

[12] The average incremental cost for a REC was calculated by averaging $806 per day and $7,486 per day (2008 dollars). While the average estimated cost per day is presented here, it is likely that the cost that is paid by a well operator will be the low incremental cost if key pieces of equipment are already present onsite or the high incremental cost if this equipment is not present onsite, and not the average of these two estimates.

5.1.4 Prevalence of Use at Oil Wells

The UT Austin study found that some co-producing oil wells are conducting RECs. It is the EPA's understanding that in some cases RECs are currently used on co-producing oil wells if pipeline infrastructure is available.

5.2 Completion Combustion Devices

5.2.1 Description

Completion combustion is a high-temperature oxidation process used to burn combustible components, mostly hydrocarbons, found in gas streams (U.S. EPA, 1991). Completion combustion devices are used to control VOC in many industrial settings, since the completion combustion devices can normally handle fluctuations in concentration, flow rate, heating value, and inert species content (U.S. EPA, Flares). These devices can be as simple as a pipe with a basic ignition mechanism and discharge over a pit near the wellhead. However, the flow directed to a completion combustion device may or may not be combustible depending on the inert gas composition of flowback gas, which would require a continuous ignition source. Completion combustion devices provide a means of minimizing vented gas during a well completion and are generally preferable to venting, due to reduced air emissions.

5.2.2 Effectiveness

Completion combustion devices can be expected to achieve 95% emission reduction efficiency, on average, over the duration of the completion or recompletion. If the energy content of natural gas is low, then the combustion mechanism can be extinguished by the flowback gas. Therefore, it may be more reliable to install an igniter fueled by a consistent and continuous ignition source. This scenario would be especially true for energized fractures where the initial flowback concentration will be extremely high in inert gases. If a completion combustion device has a continuous ignition source with an independent external fuel supply, then it is assumed to achieve an average of 95% control over the entire flowback period (U.S. EPA, 2012b).

5.2.3 Cost

An analysis of costs provided by industry for enclosed combustors was conducted by the EPA for the FBIR FIP. In addition, the State of Colorado recently completed an analysis of industry provided combustor cost data and updated their cost estimates for enclosed combustors (CDPE, 2013). Table 5-1 summarizes the data provided from each of the sources with the average cost for an enclosed combustor across these sources being $18,092. It is assumed that the cost of a continuous ignition source is included in the combustion completion device cost estimations. Also noted in the table is the most recent combustor cost used for reconsideration of control options for storage vessels under subpart OOOO.

As with RECs, because completion combustion devices are purchased for these one-time events, annual costs were assumed to be equal to the capital costs. However, multiple completions can be controlled with the same completion combustion device, not only for the lifetime of the combustion device but within the same yearly time period. Costs were estimated as the total cost of the completion combustion device itself, which corresponds to the assumption that only one device will control one completion per year. This approach may overestimate the true cost of combustion devices per well completion or recompletion.

5.2.4 Prevalence of Use at Oil Wells

The UT Austin study found that some co-producing oil wells are using completion combustion devices to reduce emissions. It is the EPA's understanding that the most common approach to reducing emissions from hydraulically fractured oil well completions is the use of a completion combustion device.

Table 5-1. Analysis of Industry Provided Enclosed Combustor Cost

Cost Parameter	Industry Provided Data					EPA Estimate in Subpart OOOO	
		FBIR			CDPHE		CDHPE
	EOG	XTO	Enerplus	QEP			
						Average of quotes	Original Data Used[a]	Adjusted Data Used[a]
Annualized Capital Cost	$5,268	$6,727	$6,116	$6,763	$3,569	$6,281	$3,546	$4,746
Other Annual Costs								
Pilot Fuel	NR	NR	NR	NR	$636		$2,078	$2,144
Operating Labor (includes management)	NR	NR	NR	NR	$10,670		$10,670	$11,012
Maintenance	NR	NR	NR	NR	$2,206			$2,190	$2,260
Data Management	NR	NR	NR	NR	$1,000		$1,095	$1,130
Total Other Annual Costs (combustor)[c]	$1,500	$23,250	$6,289	$8,500	$14,512	$10,810	$16,033	$16,546
Other Annual Costs (continuous pilot)[c]	$1,000	NR	NR	NR	included in combustor costs[b]	$1,000	included in combustor costs[b]	included in combustor costs[c]
Total Annual Costs	**$7,768**	**$29,977**	**$12,405**	**$15,263**	**$18,081**	**$18,092**	**$19,580**	**$21,292**

NR = Not reported, FBIR = Fort Berthold Indian Reservation, CDPHE = Colorado Department of Public Health and Environment, EOG = EOG Resources, XTO = XTO Energy Inc. , QEP = QEP Energy Co

a - Cost data for 40 CFR part 60, subpart OOOO updated to reflect more current cost year and equipment life (industry comments indicated a 10-year equipment life as opposed to 15 years)

b - Data used for subpart OOOO included a cost for an auto ignition system, surveillance system, VRU system, and freight and installation

c - Quotes received for FBIR FIP did not specify what was included in other annual costs.

Cost data in 2012 dollars

29

Several types of alternative use technologies are being investigated both by industry and regulators for use of associated gas.

The most prominent alternative technologies being investigated to address associated gas are liquefaction of natural gas, NGL recovery, gas reinjection, and electricity generation.

According to the Schlumberger Oilfield Glossary, "liquefied natural gas refers to natural gas, mainly methane and ethane, which has been liquefied at cryogenic temperatures. This process occurs at an extremely low temperature and a pressure near the atmospheric pressure. When a gas pipeline is not available to transport gas to a marketplace, such as in a jungle or certain remote regions offshore, the gas may be chilled and converted to liquefied natural gas (a liquid) to transport and sell it. The term is commonly abbreviated as LNG." Research is being conducted on the economic and technical feasibility of liquefaction of natural gas as a means to realize the full potential of the U.S. natural gas resources, particularly with respect to the potential of U.S. exports of LNG. However, available information indicates that this technology is typically implemented on a macro scale, requiring installation of large facilities and transportation infrastructure. Because the EPA is unaware of existing studies or further information on liquefaction of gas at the wellhead, liquefaction of natural gas is not discussed further in this paper.

Cost information is summarized to the extent that this information is readily available. In many cases, available literature does not provide cost information as the economics of the technology are still being researched.

5.3.1   Natural Gas Liquids (NGL) Recovery

Natural gas liquids are defined as "components of natural gas that are liquid at surface in field facilities or in gas-processing plants. Natural gas liquids can be classified according to their vapor pressures as low (condensate), intermediate (natural gasoline) and high (liquefied petroleum gas) vapor pressure. Natural gas liquids include propane, butane, pentane, hexane and heptane, but not methane and ethane, since these hydrocarbons need refrigeration to be liquefied. The term is commonly abbreviated as NGL."[13]

Associated gas from the Bakken formation has been termed "rich" gas, which is defined as naturally containing heavier hydrocarbons than a "lean" gas. Its liquid content adds important economic value to developments containing this type of fluid. Therefore, the value of the NGLs in the associated gas from the Bakken formation has been the subject of several studies, particularly with the concerns raised based by the rapid development of Bakken and increased flaring of associated gas. As would be expected, most of the recent studies related to NGL recovery are based on the Bakken formation.

One of these studies is the "End-Use Technology Study – An Assessment of Alternative Uses for Associated Gas" conducted by the Energy & Environmental Research Center (EERC) of the University of North Dakota (EERC, 2013). The study was conducted based on associated gas production in December 2011 and was published in 2012. This study provides an evaluation of alternative technologies and their associated costs and benefits. In particular, the study looks at NGL recovery, as a standalone operation for both recovery of salable NGLs and as a pretreatment of the associated gas for use in other local operations such as power generation.

To understand NGL recovery, the typical natural gas processing that occurs at or near the wellhead will be reviewed. Liquids and condensates (water and oil) are separated from the "wet" gas. The condensates are transported via truck or pipeline for further processing at a refinery or gas processing plant. The minimally processed wellhead natural gas is then transported to a gas-processing plant via pipeline. There, the gas is processed to remove more water, separate out NGL, and remove sulfur and carbon dioxide in preparation for release to the sales distribution system. Figure 5-1 summarizes generalized natural gas processing.

---

[13] From Schlumberger Oilfield Glossary available at
http://www.glossary.oilfield.slb.com/en/Terms.aspx?LookIn=term%20name&filter=natural+gas++liquids

Figure 5-1. Generalized Natural Gas Processing Schematic

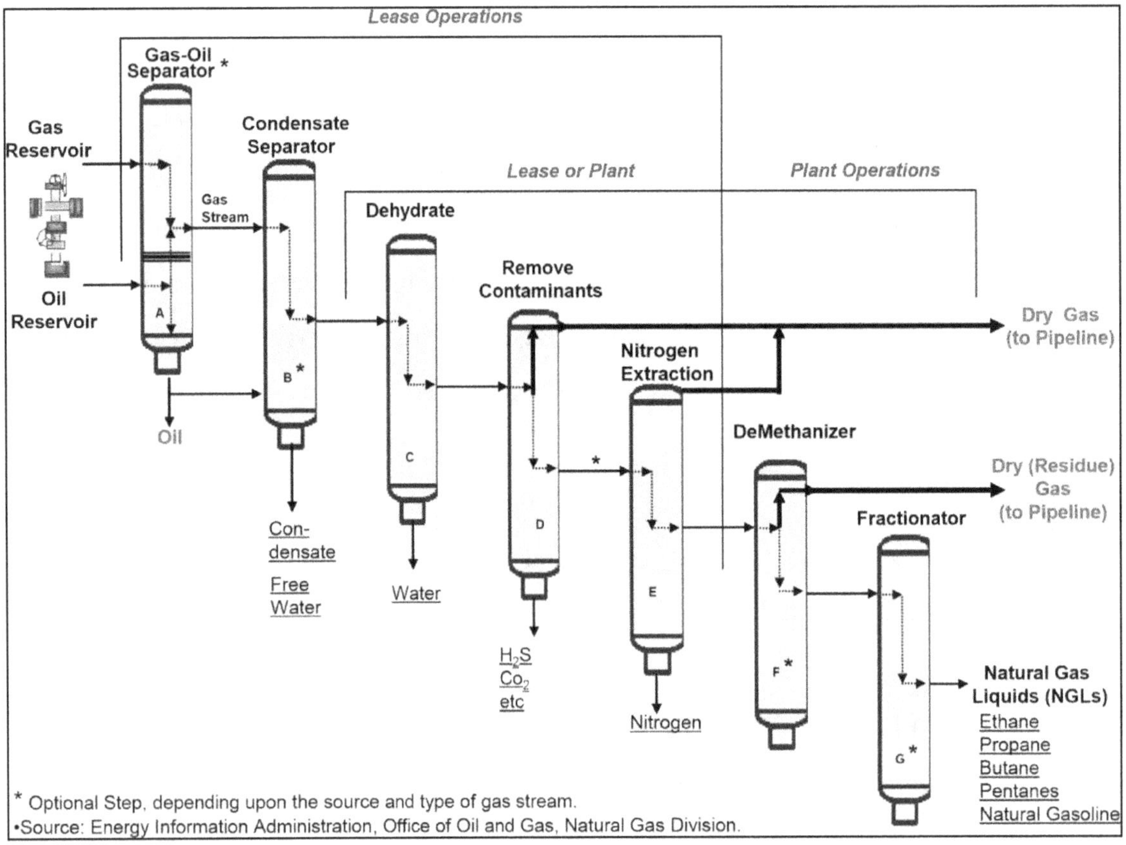

Source: U.S. EIA, 2006.

Because of the relatively high value of NGL products produced, recovery technologies have been developed both for large and small scale gas-processing applications. There are generally three approaches used in these technologies:

- Control of temperature and pressure to achieve condensation of NGLs
- Separation of heavier NGLs from lighter gas with pressurized membrane separation systems
- Physical/chemical adsorption/absorption

The typical NGL recovery technologies used are turboexpander with demethanizer, Joule-Thomson (JT) low pressure separation membranes, absorption (Refrigerated Lean Oil Separation, RLOS), adsorption using active carbon or molecular sieve, and Twister Supersonic Gas Low Temperature Separation Dew Pointing Process. For the purposes of this paper, the specifics of these technologies are not discussed; rather, the focus will be on the overall outcome and potential costs for small scale implementation at the well head for addressing associated gas.

The EERC study included a case study for a small scale NGL Recovery process at a well head. The case study evaluated the potential for deploying small scale NGL recovery systems as an interim practice to flaring associated gas while gathering lines and infrastructure were being installed or upgraded. These systems would allow the most valuable hydrocarbon portion of the gas to be captured and marketed. The leaner gas resulting could be used onsite for power generation or transported as a compressed gas. Alternatively, the leaner gas could continue to be flared. Figure 5-2 depicts the NGL Removal system flow diagram.

Figure 5-2. Natural Gas Liquids (NGL) Removal System Flow Diagram

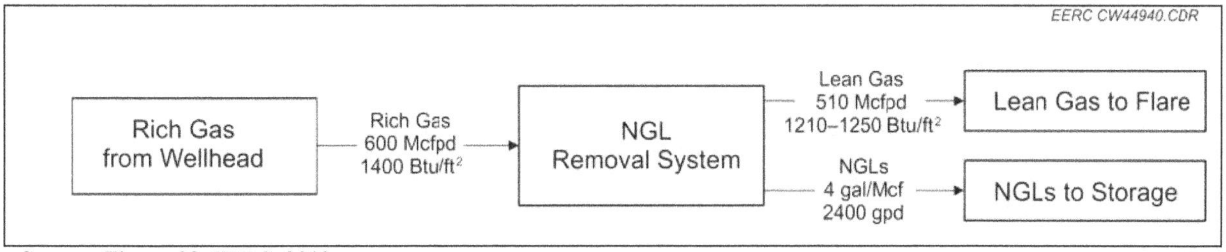

Source: **Figure 22,** EERC, 2013

According to the EERC study, 10 to 12 gallons of NGL/Mcf of associated gas is present in many producing Bakken wells. At an estimated NGL removal rate of 4 gallons/Mcf (from 1000 Mcf/day of rich gas), the daily production of NGLs would be approximately 4,000 gallons of NGLs per day (EERC, 2013). The study also states that at least at the current natural gas price, the NGLs make up a majority of the economic value of the rich gas. An evaluation of a simplified model on small-scale NGL recovery was developed based on a JT-based technology. The NGL removal system evaluation assumes the parameters shown in Table 5-2.

Table 5-2. Assumptions for NGL Recovery Case (Table 9, EERC, 2013)

Parameter	Assumed Value
Rich Gas Flow Rate from the Wellhead, average	300 Mcf/day
Rich Gas Flow Rate Processed, economic cutoff	600 Mcf/day
Rich Gas Flow Rate, design flow	1000 Mcf/day
Rich Gas Heat Content	1400 Btu/ft^3
Rich Gas Price (cost) at the Wellhead	$0.00/Mcf
Volume of NGLs Existing in Rich Gas	10–12 gallons/Mcf
NGL Price, value	$1.00/gallon
Lean Gas Flow Rate from NGL Removal System	85% of rich gas flow rate
Lean Gas Heat Content	1210–1250 Btu/ft^3
Lean Gas Price, value	$2.00/Mcf

The EERC study estimated capital and annual costs for the NGL removal system. Operating and maintenance (O&M) costs were assumed to be 10% of the total capital cost. Revenue calculations were based on NGL sales only at $1/gallon and a recovery rate of 4 gallons/Mcf. In this scenario, it has been assumed that residue gas is flared (EERC, 2013). Table 5-3, derived from Table 10 of the study, summaries the cost for the small sale NGL recovery system.

Table 5-3. Summary of NGL Removal System Costs (Table 10, EERC)

Description	Capital Cost	Annual O&M Cost
NGL Removal System, 300 Mcfd rich gas	$2,500,000	$250,000
NGL Removal System, 600 Mcfd rich gas	$2,500,000	$250,000
NGL Removal System, 1000 Mcfd rich gas	$2,500,000	$250,000

Mcfd = One thousand standard cubic feet per day.

The EERC study concluded that the technical aspects of NGL recovery are fairly straight forward; however, the business aspects are much more complicated, particularly with respect to NGL product supply chain and contractual considerations. Further, the study concluded that NGL recovery would be most economical at wells flaring larger quantities of gas immediately after production begins. Other attributes that would be important for the economic feasibility of the NGL recovery system would be that the systems are mobile and easily mobilized, and that infrastructure with respect to truing of NGL production is available.

5.3.2   Natural Gas Reinjection

Schlumberger's Oilfield Glossary defines gas injection as "a reservoir maintenance or secondary recovery method that uses injected gas to supplement the pressure in an oil reservoir or field. In most cases, a field will incorporate a planned distribution of gas-injection wells to maintain reservoir pressure and effect an efficient sweep of recoverable liquids."[14]

The industry has employed production methods to increase production, which are termed enhanced oil recovery (EOR) or improved oil recovery (IOR) (Rigzone, 2014). These methods are generally considered to be tertiary methods employed after waterflooding or pressure maintenance. The practice involves injecting gas into the gas cap of the formation and boosting the depleted pressure in the formation with systematically placed injection wells throughout the field. The pressure maintenance methods maybe employed at the start of production or introduced after the production has started to lessen. The reinjection of natural gas is the use of associated gas at the same oilfield to accomplish the goals of gas injection as defined above. The increase in the pressure within the reservoir helps to induce the flow of crude oil. After the crude has been pumped out, the natural gas is once again recovered.

Natural gas injection is also referred to as cycling. Cycling is used to prevent condensate from separating from the dry gas in the reservoir due to a drop in reservoir pressure. The condensate liquids block the pores within the reservoir, making extraction practically impossible. The NGL are stripped from the gas on the surface after it has been produced, and the dry gas is

---

[14] Schlumberger Oilfield Glossary, available at
http://www.glossary.oilfield.slb.com/en/Terms.aspx?LookIn=term%20name&filter=gas+injection

then re-injected into the reservoirs through injection wells. Again, this helps to maintain pressure in the reservoir while also preventing the separation within the hydrocarbon (Rigzone, 2014). Figure 5-3 illustrates the relationship between the gas injection well and the production well.

Figure 5-3. Gas Injection and Production Well

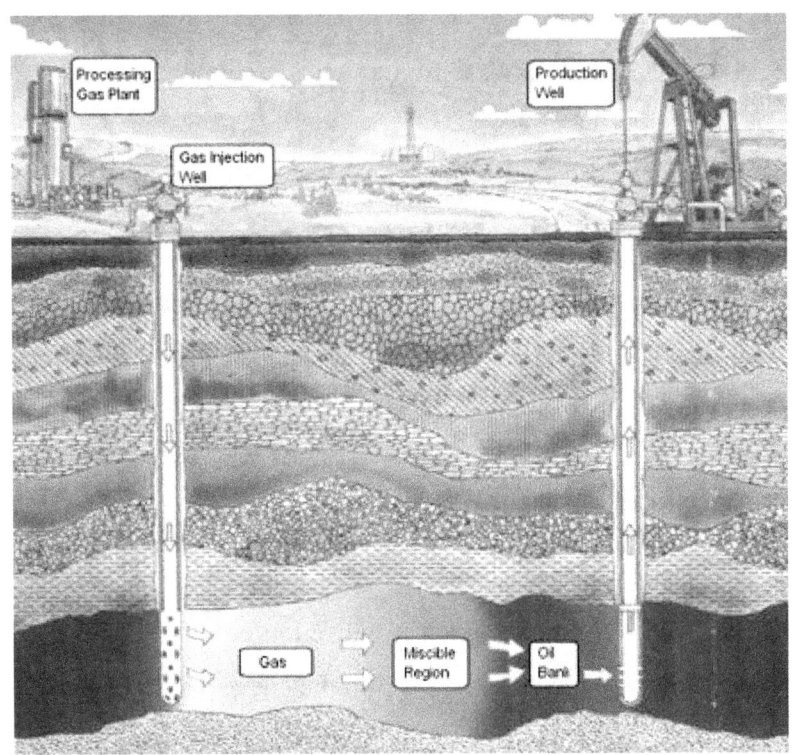

Source: Rigzone, 2014

In the scenarios that were found in available literature, the dry gas is also used as fuel onsite for the generators that power the reinjection pumps. Therefore, the costs associated with the process are mainly initial capital costs. No published information was obtained on the capital and annual costs for these operations.

Figure 5-4 presents a fully implemented gas injection project scheme. In this scheme, associated gas from an oil well (or natural gas from a gas well) is processed through a gas cycling facility (GCF) where recoverable NGLs are separated from methane and the resulting methane is either used for onsite power generation or re-injected in to the formation.

Figure 5-4. Gas Cycling Facility Project Flow

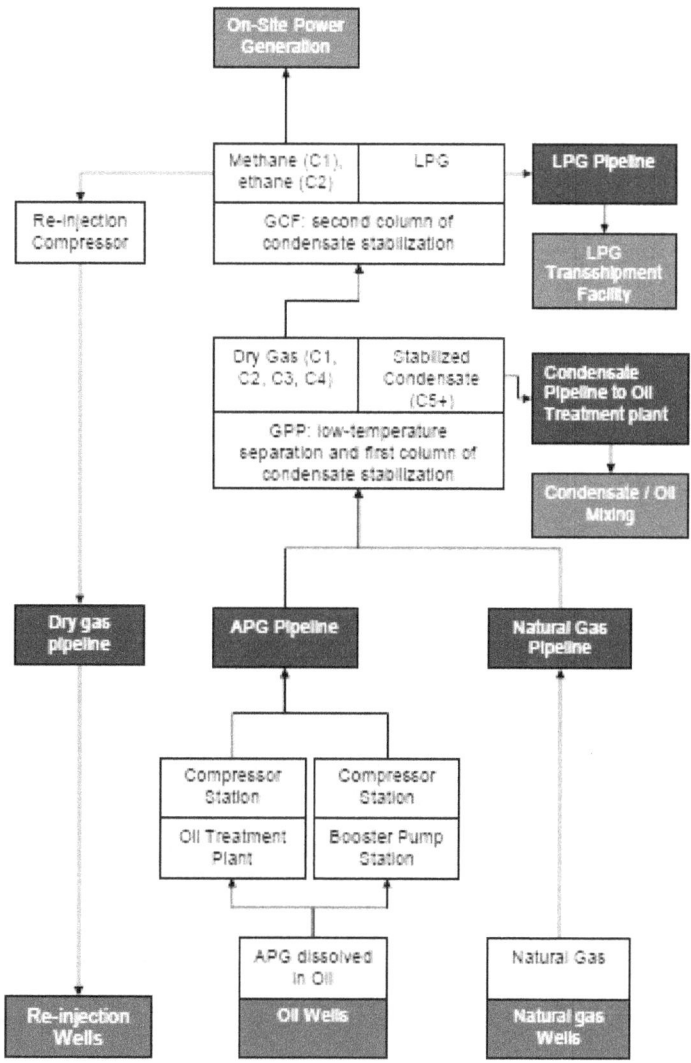

The literature that was reviewed evaluated gas reinjection projects only from the perspective of an enhanced oil recovery opportunity and did not specifically discuss the quantity or percentage of associated gas emissions that were eliminated through the process. The EPA is not aware of literature that discusses the efficacy of mitigating associated gas emissions using the natural gas reinjection process. The efficacy would be highly dependent on many factors, which include the composition value of the gas and the availability of transmission infrastructure. Further, because the use of this process to reduce associated gas emissions in conjunction with oil recovery is an emerging technology, the prevalence of use in the industry and estimated cost to implement the process is unknown to the EPA.

5.5.3   Electricity Generation for Use Onsite

As discussed above, associated gas can be used for generation of electrical power to be used onsite. The EERC study stated that power generation technologies would need to be designed to match the variable wellhead gas flow rates and gas quality, and would need to be constructed for mobility. The EERC study discussed previously also looked at options for use of associated gas for power generation. The EERC study included an evaluation of several technologies fired by natural gas both for grid support (i.e., power generation for direct delivery onto the electric grid) and local power (i.e., power generation for local use with excess generation, if any, sent to the electrical grid). This study provides one of the most comprehensive and recent evaluations of the economics of use of associated gas for electric generation. Therefore, the case study results of this study are used to discuss the cost of this technology for this paper.

Although grid support is potentially a viable use for this gas, it is not considered to be an emissions reduction technology for the purposes of this paper. Grid support requires an infrastructure similar in scope as that needed to bring gas to market. The focus of this section of the paper is on the venting or flaring of associated gas due to the lack of infrastructure to bring it to market. It is unlikely that a well site that is lacking pipeline infrastructure would have access to the necessary infrastructure to provide grid support. Therefore, the focus here is on the use of the gas at the local level, either directly at the wellpad or in an immediate oilfield region to support local activities. The benefits of using associated gas to provide electricity for these activities are both reducing the quantity of gas vented and reducing the quantity of other types of fuel used (e.g., diesel).

The EERC study considered a local power project to be wellhead gas (with limited cleanup) being piped to an electrical generator that produces electricity which is first used to power local consumption (e.g., well pad, group of wells, or an oilfield) with any excess electricity put on the electrical grid for distribution by the local utility to its customers. These projects can range widely in scale, depending on the goal of the project (i.e., satisfy only local load, satisfy local load with minimal excess generation, or satisfy local load with significant excess generation). The study evaluated two power generation scenarios: reciprocating engine and a microturbine.

The first step in using associated gas for electric generation is removal of NGLs from the rich gas. Removal of the NGLs significantly increases the performance of the genset and reduces the loss of resource (when flaring is necessary). According to the EERC study, removal of NGLs such as butane and some propane could be accomplished using a low temperature separation process. The study found that small, modular configurations of these types of systems are not widely available. The estimated capital cost for the NGL removal and storage system is $2,500,000. This capital cost includes the necessary compression to take the rich gas from the heater/treater at 35 psi up to 200 - 1000 psi delivered to the NGL removal system as well as the cost for four 400-bbl NGL storage tanks (EERC, 2013). The study authors considered NGL recovery a valuable first step; however, they also stated that it was not necessary in all circumstances.

The study made certain assumptions about the flow of associated gas from the wellhead and fuel consumption of the respective electrical generator for the case study. Table 5-4 summarizes the assumed wellhead gas flow for the case study. Figure 5-5 shows a block flow diagram of an example NGL removal system.

Table 5-4. Summary of Wellhead Gas Flow and Product Volume Assumptions

Scenario	Rich Gas Flow, Mcf/day	NGLs Produced, gallons/day	Lean Gas Produced, Mcf/day
Reciprocating Engine	600	2,400	510
Microturbine	600	2,400	510

Source Table 33, EERC 2013

Figure 5-5. NGL Removal System Block Flow Diagram

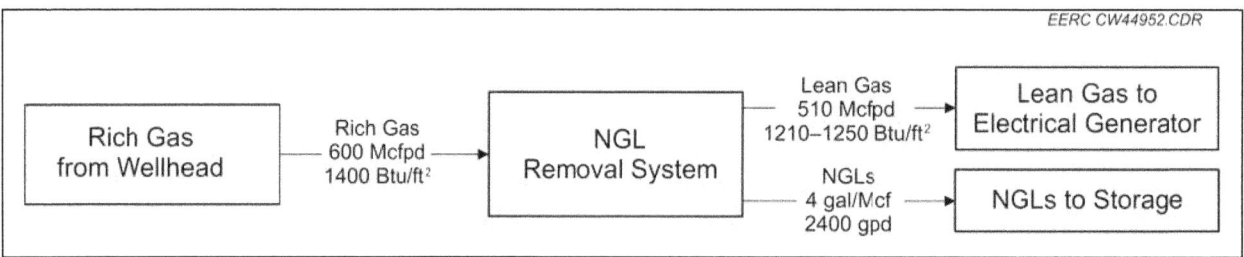

For the case study, the authors targeted a power production scenario of 1 MW for the reciprocating engine and 200 kW for the microturbine. Both scenarios used the same NGL removal system prior to introduction of the rich gas to the generator. Figure 5-6 depicts the process flow diagram for the local power generation scenario.

Figure 5-6. Process Flow Diagram, Local Power Generation Scenario

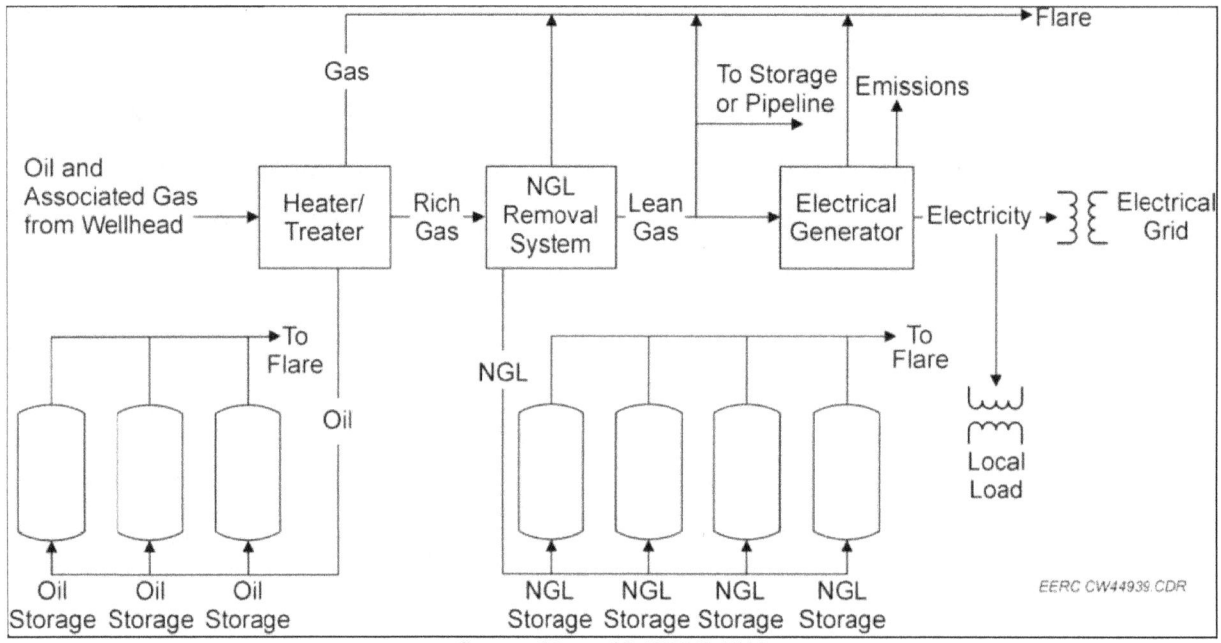

For the reciprocating engine scenario, vendor provided costs for a 250-kW natural gas fired reciprocating engine genset was $200,000. The study estimated the annual O&M cost was assumed to be 10% of the capital cost. The costs for this scenario are summarized in Table 5-5.

Table 5-5. Total Cost Summary - Reciprocating Engine Scenario

	Capital Cost	Annual Cost
NGL Removal and Storage System	$2,500,000	$250,000
Electrical Generator System	$200,000	$20,000
All Other Infrastructure	$500,000	
Total Capital Cost	**$3,200,000**	**$270,000**

Source Table 38, EERC 2013

For the microturbine scenario, the authors chose to analyze a four, 65 kW microturbine package rated to provide approximately 195 kW of power. This scenario also involved the removal of NGLs prior to delivery of gas to the microturbine and the use of generated electricity to satisfy local electrical demand, with the excess electricity delivered to the grid. The authors noted that the volume of gas generated from the wellhead(s) will determine the size of the system needed and that a range of generation scales should be considered for optimum performance. The process flow for this scenario is the same as shown above in Figure 5-6.

The NGL removal system is likely to be much larger in processing capacity than the electrical generation system. Generally, the NGL removal system will be most economical only at the higher-gas-producing wells. The microturbine package evaluated consumed less than 100 Mcf/day, which meant that excess gas would either need to be flared or the project must be designed to store the excess gas for sale to the pipeline. In the scenarios described here, the authors assumed that the excess lean gas is sold.

For the microturbine system analyzed, the vendors offered a factory protection plan (FPP) that covers all scheduled and unscheduled maintenance of the system as well as parts, including an overhaul or turbine replacement at 40,000 hours of operation. The FPP "locks in" the annual O&M cost of the system and, in both scenarios presented below, it is assumed that the FPP is purchased (EERC, 2013). Table 5-6 summarizes the capital and annual O&M costs for the microturbine system, as well as the NGL recovery system discussed above.

Table 5-6. Total Cost Summary - Microturbine Scenario (Four 65-kW)

	Capital Cost	Annual Cost
NGL Removal and Storage System	$2,500,000	$250,000
Electrical Generator System	$383,200	$33,640
All Other Infrastructure	$500,000	
Total Capital Cost	**$3,382,200**	**$283,640**

Source: Table 41, EERC, 2013.

The study authors also evaluated revenue potential for electricity sent to the grid as an offset to the costs summarized above. Their analysis indicated that based on cost (discussed above) and their revenue assumptions, both scenarios provided a simple payback of 3 years or less. However, given the substantial upfront capital costs of these options, these options may not be preferable to building the necessary pipeline infrastructure to take the gas to market.

In addition to the electric generation potential for associated gas, the study also discussed the use of wellhead gas as a fuel for drilling operations. The authors indicated that the EERC is working with Continental Resources, ECO-AFS, Altronics, and Butler Caterpillar to conduct a detailed study and field demonstration of the GTI Bi-Fuel System. Within that task, the EERC conducted a series of tests at the EERC using a simulated Bakken gas designed to test the operational limits of fuel quality and diesel fuel replacement while monitoring engine performance and emissions. The authors indicated that the Bi-Fuel System is an aftermarket addition to the system allowing natural gas to the air intake, and the engine performance is unaltered from the diesel operation. This system, as the name implies, could be used on either fuel without requiring any alterations.

According to the study report, total installed capital cost for the Bi-Fuel System ranges from $200,000 to $300,000 (EERC, 2013). Other costs that would be incurred would be those for piping wellhead gas to the engine building. The study did not include those costs because they can be highly variable depending on the distance to the nearest gas source and gas lease terms.

The study reports that ECO-AFS had recently installed several Bi-Fuel Systems on rigs in the Williston Basin and that early data suggest that diesel fuel savings of approximately $1 to $1.5 million can be achieved annually. Under typical conditions, operators can expect to achieve diesel replacement of 40% - 60% at optimal engine loads of 40% - 50% (EERC, 2013).

The EERC study noted that there are a number of other potential natural gas uses related to oil production and operations that could take advantage of rich gas on a well site. Those would include:

- Heating of drilling fluids during winter months (replacing the diesel or propane fuel used currently)
- Providing power for hydraulic fracturing operations decreasing reliance on diesel fuel (i.e., by using Bi-fuel systems)
- Providing fuel for workover rigs (if the rig is equipped with a separate generator)

6.0 SUMMARY

As discussed in the previous sections, the EPA used the body of knowledge presented in this paper to summarize its understanding of emissions characterization and potential emissions mitigation techniques for oil well completions and associated gas. From that body of knowledge, the following statements summarize the EPA's understanding of the state of the industry with respect to these sources of emissions:

- Available estimates of uncontrolled emissions from hydraulically fractured oil well completions are presented below:

Study	Average Uncontrolled VOC Emissions (Tons/Completion)	Average Uncontrolled Methane Emissions (Tons/Completion)
Fort Berthold Federal Implementation Plan	37	N/A
ERG/ECR Analysis of HPDI® Data (7 day flowback period)	20.2	24
ERG/ECR Analysis of HPDI® Data (3 day flowback period)	6.4	7.7
EDF/Stratus Analysis of HPDI® Data (Eagle Ford)	N/A	27.2
EDF/Stratus Analysis of HPDI® Data (Wattenberg)	N/A	10.5
EDF/Stratus Analysis of HPDI® Data (Bakken)	N/A	19.8
Measurements of Methane Emissions at Natural Gas Production Sites in the United States	N/A	213
Methane Leaks from North American Natural Gas Systems (Eagle Ford)	N/A	90.9
Methane Leaks from North American Natural Gas Systems (Bakken)	N/A	31.1
Methane Leaks from North American Natural Gas Systems (Permian)	N/A	31.2

- Limited information is available on uncontrolled emissions from hydraulically fractured oil well recompletions, and controlled emission factors for hydraulically fractured oil well completions and recompletions.

- National level estimates of uncontrolled methane emissions from hydraulically fractured oil well completions range from 44,306 tons per year (ERG/ECR) to 247,000 tons per year (EDF/Stratus analysis).

- One study (ERG/ECR) estimated nationwide uncontrolled VOC emissions from hydraulically fractured oil well completion to be 116,230 tons per year assuming a 7-day flowback period and 36,825 tons per year assuming a 3-day flowback period.

- There is some data that shows (Allen et. al.) that RECs, in certain situations, can be an effective emissions control technique for oil well completions when gas is co-produced.

However, there may be a combination of well pressure and gas content below which RECs are not technically feasible at co-producing oil wells.

- Some oil well completions are controlled using RECs; however, national data on the number of completions that are controlled using a REC are not available. It is the EPA's understanding that most oil well completion emissions are controlled with combustion; however, data on an average percentage are not available. Likewise, data are not available on the percentage of oil wells nation-wide that vent completion emissions to the atmosphere.

- Other gas conserving technologies are being investigated for use in completions and for control of associated gas emissions. These include gas reinjection, NGL recovery and use of the gas for power generation for local use. Some studies have evaluated the economics of some of these technologies and determined, in some cases, they can result in net savings to the operator depending on the value of the recovered gas or liquids or the value of the power generated. However, some barriers exist with respect to technology availability and application of the technology to varying scales of oil well gas production. In addition, costs vary for implementing some of these technologies.

7.0 CHARGE QUESTIONS FOR REVIEWERS

1. Please comment on the national estimates and per well estimates of methane and VOC emissions from hydraulically fractured oil well completions presented in this paper. Are there factors that influence emissions from hydraulically fractured oil well completions that were not discussed in this paper?

2. Most available information on national and per well estimates of emissions is on uncontrolled emissions. What information is available for emissions, or what methods can be used to estimate net emissions from uncontrolled emissions data, at a national and/or at a per well level?

3. Are further sources of information available on VOC or methane emissions from hydraulically fractured oil well completions beyond those described in this paper?

4. Please comment on the various approaches to estimating completion emissions from hydraulically fractured oil wells in this paper.

  - Is it appropriate to estimate average uncontrolled oil well completion emissions by using the annual average daily gas production during the first year and multiplying that value by the duration of the average flowback period?

  - Is it appropriate to estimate average uncontrolled oil well completion emissions using "Initial Gas Production," as reported in DI Desktop, and multiplying by the flowback period?

  - Is it appropriate to estimate average uncontrolled oil well completion emissions by increasing emissions linearly over the first nine days until the peak rate is reached (normally estimated using the production during the first month converted to a daily rate of production)?

  - Is the use of a 3-day or 7-day flowback period for hydraulically fractured oil wells appropriate?

5. Please discuss other methodologies or data sources that you believe would be appropriate for estimating hydraulically fractured oil well completion emissions.

6. Please comment on the methodologies and data sources that you believe would be appropriate to estimate the rate of recompletions of hydraulically fractured oil wells. Can data on recompletions be used that does not differentiate between conventional oil wells and hydraulically fractured oil wells be reasonably used to estimate this rate? For example, in the GHG Inventory, a workover rate of 6.5% is applied to all oil wells to estimate the number of workovers in a given year, and in the ERG/ECR analysis above a rate of 0.5% is developed based on both wells with and without hydraulic fracturing. Would these rates apply to hydraulically fractured oil wells? For hydraulically fractured gas wells, the GHG Inventory uses a refracture rate of 1%. Would this rate be appropriate for hydraulically fractured oil wells?

7. Please comment on the feasibility of the use of RECs or completion combustion devices during hydraulically fractured oil well completion operations. Please be specific to the types of wells where these technologies or processes are feasible. Some characteristics that should be considered in your comments are well pressure, gas content of flowback, gas to oil ratio

(GOR) of the well, and access to infrastructure. If there are additional factors, please discuss those. For example, the Colorado Oil and Gas Conservation Commission requires RECs only on "oil and gas wells where reservoir pressure, formation productivity and wellbore conditions are likely to enable the well to be capable of naturally flowing hydrocarbon gas in flammable or greater concentrations at a stabilized rate in excess of five hundred (500) MCFD to the surface against an induced surface backpressure of five hundred (500) psig or sales line pressure, whichever is greater."[15]

8. Please comment on the costs for the use of RECs or completion combustion devices to control emissions from hydraulically fractured oil well completions.

9. Please comment on the emission reductions that RECs and completion combustion devices achieve when used to control emissions from hydraulically fractured oil well completions.

10. Please comment on the prevalence of the use of RECs or completion combustion devices during hydraulically fractured oil well completion and recompletion operations. Are you aware of any data sources that would enable estimating the prevalence of these technologies nationally?

11. Did the EPA correctly identify all the available technologies for reducing gas emissions from hydraulically fractured oil well completions or are there others?

12. Please comment on estimates of associated gas emissions in this paper, and on other available information that would enable estimation of associated gas emissions from hydraulically fractured oil wells at the national- and the well-level.

13. Please comment on availability of pipeline infrastructure in hydraulically fractured oil formations. Do all tight oil plays (e.g., the Permian Basin and the Denver-Julesberg Basin) have a similar lack of infrastructure that results in the flaring or venting of associated gas?

14. Did the EPA correctly identify all the available technologies for reducing associated gas emissions from hydraulically fractured oil wells or are there others? Please comment on the

---

[15] Colorado Department of Natural Resources: Oil and Gas Conservation Commission Rule 805.b(3)A. (http://cogcc.state.co.us/)

costs of these technologies when used for controlling associated gas emissions from hydraulically fractured oil wells. Please comment on the emissions reductions achieved when these technologies are used for controlling associated gas emissions from hydraulically fractured oil wells.

15. Are there ongoing or planned studies that will substantially improve the current understanding of VOC and methane emissions from hydraulically fractured oil well completions and associated gas and available options for increased product recovery and emission reductions?

8.0   REFERENCES

Booz Allen Hamilton, Wyoming Heritage Foundation. 2008. *Oil and Gas Economic Contribution Study*. August 2008. Available at http://www.westernenergyalliance.org/wp-content/uploads/2009/05/WYHF_O_G_Economic_Study_FINAL.pdf.

Brandt, A.R., et al. 2014a. *Methane Leaks from North American Natural Gas Systems.* Science 343, 733 (2014). February 14, 2014. Available at http://www.novim.org/images/pdf/ScienceMethane.02.14.14.pdf.

Brandt, A.R., et al. 2014b. *Supplementary Materials for Methane Leaks from North American Natural Gas Systems*. Science 343, 733 (2014). February 14, 2014. Available at http://www.novim.org/images/pdf/ScienceSupplement.02.14.14.pdf

Canadian Association of Petroleum Producers (CAAP). 2004. *A National Inventory of Greenhouse Gas (GHSP Criteria Air Contaminant (CAC) and Hydrogen Sulfide H_2S) Emission by the Upstream Oil and Gas Industry*. CAPP Pub. No. 2005-0013.

Ceres, 2013. *Flaring Up: North Dakota Natural Gas Flaring More Than Doubles in Two Years.* Salman, Ryan and Logan, Andrew. July 2013.

Colorado Department of Public Health and Environment (CDPE). 2013. APCD 2013 Rulemaking April Stakeholder Meeting, Presentation. April 25, 2013.

EC/R, Incorporated. 2010a. Memorandum to Bruce Moore from Denise Grubert. *American Petroleum Institute Meeting Minutes*. EC/R, Incorporated. July 2010.

EC/R, Incorporated. 2010b. Memorandum to Bruce Moore from Denise Grubert. *SHWEP Site Visit Report*. EC/R Incorporated. November 2010.

EC/R, Incorporated. 2011a. Memorandum to Bruce Moore from Heather Brown. *Composition of Natural Gas for Use in the Oil and Natural Gas Sector Rulemaking*. EC/R, Incorporated. June 29, 2011.

EC/R, Incorporated. 2011b. Memorandum to Bruce Moore from Denise Grubert. *American Petroleum Institute Meeting Minutes Attachment 1: Review of Federal Air Regulations for the Oil and Natural Gas Sector 40 CFR Part 60, Subparts KKK and LLL; 40 CFR Part 63 Subparts HH and HHH*. EC/R, Incorporated. February 2011.

Energy & Environmental Research Center (EERC). 2013. *End-Use Technology Study – An Assessment of Alternative Uses for Associated Gas*. EERC Center for Oil and Gas, University of North Dakota, Grand Forks, ND.

Environmental Defense Fund (EDF). 2014. *Co-Producing Wells as a Major Source of Methane Emissions: A Review of Recent Analyses, March, 2014*. Available at http://blogs.edf.org/energyexchange/files/2014/03/EDF-Co-producing-Wells-Whitepaper.pdf. Supplemental materials available at https://www.dropbox.com/s/osrom4w6ewow4ua/EDF-Initial-Production-Cost-Effectiveness-Analysis.xlsx.

Environmental Research Group, Inc. (ERG). 2013. *Hydraulically Fractured Oil Well Completions*. October 24, 2013 (available as Appendix A).

ICF, International (ICF). 2011. Memorandum to Bruce Moore from ICF Consulting. *Percent of Emissions Recovered by Reduced Emission Completions*. ICF Consulting. May 2011.

North Dakota Pipeline Authority (NDPA). 2013. North Dakota Natural Gas, *A Detailed Look At Natural Gas Gathering*. October 21, 2013.

Oil and Gas Journal (OGJ). 2012. *Restricting North Dakota gas-flaring would delay oil output, impose costs*. November 5, 2012. Available at http://www.ogj.com/articles/print/vol-110/issue-11/drilling-production/restricting-north-dakota-gas-flaring-would.html.

Proceeding of the National Academy of Sciences of the United States of America (PNAS). 2013. *Measurement of Methane Emissions at Natural Gas Production Sites in the United States*. August 19, 2013. Available at http://www.pnas.org/content/early/2013/09/10/1304880110.abstract.

Railroad Commission of Texas (RRCTX). 2013. *Oil Production and Well Counts (1935-2012), History of Texas Initial Crude Oil, Annual Production and Producing Wells*. Available at http://www.rrc.state.tx.us/data/production/oilwellcounts.php; and *Summary of Drilling, Completion and Plugging Reports Processed for 2012*. January 7, 2013. Available at http://www.rrc.state.tx.us/data/drilling/drillingsummary/2012/annual2012.pdf

Rigzone. 2014. *How Does Gas Injection Work?* Available from http://www.rigzone.com/training/insight.asp?insight_id=345&c_id=4. Accessed March 2014.

State of Colorado Oil & Gas Conservation Commission. (COGCC) 2012, Staff Report. November 15, 2012 available at www.colorado.gov/cogcc.

Swindell. 2012. *Eagle Ford Shale – An Early Look at Ultimate Recovery*. SPE 158207. SPE annual Technical Conference and Exhibition held in San Antonio, Texas, USA. Available at http://gswindell.com/sp158207.pdf.

U.S. Energy Information Administration (U.S. EIA). 2010. Annual U.S. Natural Gas Wellhead Price. Energy Information Administration. Natural Gas Navigator. Retrieved December 12, 2010. Available at http://www.eia.doe.gov/dnav/ng/hist/n9190us3a.htm.

U.S. Energy Information Administration (U.S. EIA). 2011. Annual Energy Outlook 2011 and an update on EIA activities. Available at http://www.eia.gov/pressroom/presentations/newell_02082011.pdf.

U.S. Energy Information Administration (U.S. EIA). 2012a. Total Energy Annual Energy Review. Table 6.4 Natural Gas Gross Withdrawals and Natural Gas Well Productivity, Selected Years, 1960 - 2011. (http://www.eia.gov/total energy/data/annual/pdf/sec6_11.pdf).

U.S. Energy Information Administration (U.S. EIA). 2012b. Total Energy Annual Energy Review. Table 5.2 Crude Oil Production and Crude Oil Well Productivity, Selected Years, 1954 - 2011. (http://www.eia.gov/total energy/data/annual/pdf/sec5_9.pdf).

U.S. Energy Information Administration (U.S. EIA). 2013a. *Drilling often results in both oil and natural gas production*. October, 2013. Available at http://www.eia.gov/todayinenergy/detail.cfm?id=13571.

U.S. Energy Information Administration (U.S. EIA). 2013b. Annual Energy Outlook 2013. Available at http://www.eia.gov/forecasts/aeo/pdf/0383%282013%29.pdf.

U.S. Environmental Protection Agency (U.S. EPA). *Air Pollution Control Technology Fact Sheet: FLARES*. Clean Air Technology Center.

U.S. Environmental Protection Agency (U.S. EPA). 1991. *AP 42, Fifth Edition, Volume I, Chapter 13.5 Industrial Flares*. EPA/Office of Air Quality Planning & Standards. 1991.

U.S. Environmental Protection Agency (U.S. EPA). 1996. *Methane Emissions from the Natural Gas Industry Volume 2: Technical Report, Final Report*. Gas Research Institute and U.S. Environmental Protection Agency. Washington, DC. June, 1996.

U.S. Environmental Protection Agency (U.S. EPA). 1999. *U. S. Methane Emissions 1990-2020: Inventories, Projections, and Opportunities for Reductions*. Washington, DC, 1999.

U.S. Environmental Protection Agency (U.S. EPA). 2004. Fact Sheet No. 703: Green Completions. Office of Air and Radiation: Natural Gas Star Program. Washington, DC. September 2004.

U.S. Environmental Protection Agency (U.S. EPA). 2011a. *Lessons Learned: Reduced Emissions Completions for Hydraulically Fractured Gas Wells. Office of Air and Radiation: Natural Gas Star Program*. Washington, DC. 2011. Available at http://epa.gov/gasstar/documents/reduced_emissions_completions.pdf.

U.S. Environmental Protection Agency (U.S. EPA). 2011b. *Oil and Natural Gas Sector: Standards of Performance for Crude Oil and Natural Gas Production, Transmission, and Distribution. Background Technical Support Document for Proposed Standards.* July 2011. EPA-453/R-11002.

U.S. Environmental Protection Agency (U.S. EPA). 2012a. *Technical Support Document, Federal Implementation Plan for Oil and Natural Gas Well Production Facilities; Fort Berthold Indian Reservation (Mandan, Hidatsa, and Arikara Nations), North Dakota.* Docket Number: EPA-R08-OAR-2012-0479.

U.S. Environmental Protection Agency (U.S. EPA). 2012b. *Oil and Natural Gas Sector: Standards of Performance for Crude Oil and Natural Gas Production, Transmission, and Distribution. Background Supplemental Technical Support Document for Proposed Standards.* April 2012.

U.S. Environmental Protection Agency. (U.S. EPA) 2013. *Petroleum and Natural Gas Systems: 2012 Data Summary. Greenhouse Gas Reporting Program.* October 2013. (http://www.epa.gov/ghgreporting/documents/pdf/2013/documents/SubpartW-2012-Data-Summary.pdf).

U.S. Environmental Protection Agency (U.S. EPA). 2014. *Inventory of Greenhouse Gas Emissions and Sinks: 1990-2012.* Washington, DC. April 2014. (http://www.epa.gov/climatechange/Downloads/ghgemissions/US-GHG-Inventory-2014-Chapter-3-Energy.pdf).